Abdelhafid Mimouni

Antagonism of Orexine and Adenosine on Vigilance

Abdelhafid Mimouni

Antagonism of Orexine and Adenosine on Vigilance

ScienciaScripts

Imprint

Any brand names and product names mentioned in this book are subject to trademark, brand or patent protection and are trademarks or registered trademarks of their respective holders. The use of brand names, product names, common names, trade names, product descriptions etc. even without a particular marking in this work is in no way to be construed to mean that such names may be regarded as unrestricted in respect of trademark and brand protection legislation and could thus be used by anyone.

Cover image: www.ingimage.com

This book is a translation from the original published under ISBN 978-620-6-71945-8.

Publisher:
Sciencia Scripts
is a trademark of
Dodo Books Indian Ocean Ltd. and OmniScriptum S.R.L publishing group

120 High Road, East Finchley, London, N2 9ED, United Kingdom
Str. Armeneasca 28/1, office 1, Chisinau MD-2012, Republic of Moldova, Europe
Printed at: see last page
ISBN: 978-620-8-02921-0

"Antagonism of Orexine and Adenosine on Vigilance".

Author:

Dr. Abdelhafid Mimouni independent researcher specialized in bioinorganic systems chemistry. Extensive expertise in macromolecular synthesis and characterization. He obtained his PhD in chemistry from the University of Paris XII in 1997 and a Diplôme des études approfondies in bioinorganic systems from the University of Paris XI in 1993.

Summary:

This book examines the importance of orexin and adenosine in the regulation of sleep and wakefulness, exploring their molecular mechanisms, clinical implications, as well as technological advances and research prospects. It covers neuroanatomy, neurophysiology, and their interaction with other neurotransmitters. Subsequent sections focus on biosynthesis, specific receptors, and the influence of metals on their function. Clinical aspects cover disorders such as narcolepsy and existing and potential therapeutic approaches. Finally, the book discusses advanced imaging methods, experimental models, and future innovations in the field, highlighting the growing importance of these neurotransmitters in modern neurological research.

Book map

2

Introduction

Gélineau syndrome, also known as narcolepsy-cataplexy, is a chronic neurological disorder that profoundly alters the control of sleep and wakefulness in affected individuals. Characteristic symptoms include excessive daytime sleepiness, sudden and uncontrollable sleep episodes, and manifestations of cataplexy, a temporary loss of muscle tone triggered by strong emotions such as joy, surprise or anger. In addition, patients may experience hypnagogic or hypnopompic hallucinations, as well as nocturnal sleep disturbances such as insomnia. Diagnosis is based on a thorough assessment including history, physical examination and specialized tests such as polysomnography and sleep latency tests. Treatment is aimed at alleviating symptoms and may include the administration of medication to control daytime sleepiness and cataplexy, while offering tailored support to improve patients' quality of life.

Chapter 1: Neurobiological foundations

Section 1.1: Neuroanatomy and neurophysiology of Orexin and Adenosine

The regulation of sleep and wakefulness is strongly influenced by orexin and adenosine, two key neurotransmitters of the central nervous system. Orexin, also known as hypocretin, is produced in the hypothalamus and plays a central role in maintaining wakefulness and alertness. Adenosine, on the other hand, builds up progressively during wakefulness and promotes drowsiness by acting on specific receptors in the brain.

Section 1.2: Specific roles in the regulation of the sleep-wake cycle

The interaction between orexin and adenosine is crucial for the fine modulation of the sleep-wake cycle. Orexin acts in opposition to the effects of adenosine, promoting wakefulness and vigilance. Orexin deficiency, observed in patients with narcolepsy-cataplexy, leads to excessive daytime sleepiness due to an inability to maintain wakefulness adequately.

Section 1.3: Interactions with other neurotransmitters and neurological systems

In addition to their direct impact on the regulation of sleep and wakefulness, orexin and adenosine interact closely with various other neurotransmitters and neural systems to modulate their function. This complex interconnection plays a crucial role in the diversity of symptoms observed in patients suffering from sleep disorders, notably narcolepsy-cataplexy.

For example, orexin acts synergistically with neurotransmitters such as dopamine and noradrenaline, influencing the regulation of alertness and motor activity. Similarly, adenosine, in addition to its effects on specific receptors, may interact with other neural systems involved in modulating sleep and the response to environmental stresses.

This complexity of the neurochemical interactions underlying the functions of orexin and adenosine underscores the importance of understanding these mechanisms to develop more effective and targeted therapeutic strategies for the treatment of sleep disorders.

Chapter 2: Molecular and cellular mechanisms

Section 2.1: Orexin and Adenosine biosynthesis and metabolism

Understanding the processes of orexin and adenosine biosynthesis and metabolism is key to understanding their regulation and dysfunction in individuals with sleep disorders. These processes offer potential targets for the development of new therapies aimed at restoring proper regulation of the sleep-wake cycle.

To delve deeper into the section on orexin and adenosine biosynthesis and metabolism, it's essential to examine the underlying biochemical processes that govern the production and processing of these neurotransmitters in the central nervous system.

Biosynthesis and metabolism of Orexin (Hypocretin)

Orexin, also known as hypocretin, is mainly synthesized in the hypothalamus, a key brain region involved in the regulation of sleep and wakefulness. The synthesis of orexin begins with the transcription of its precursor, prepro-orexin, from the HCRT gene located on chromosome 17 in humans. This precursor is then cleaved to produce the active orexin peptides, orexin A and orexin B, by specific enzymes present in hypothalamic neurons.

Once released into the cerebrospinal circulation, orexin acts primarily by binding to two main receptors: orexin type 1 receptor (OX1R) and orexin type 2 receptor (OX2R). These receptors are expressed differently in different regions of the brain, modulating various aspects of sleep and wake physiology, including regulation of muscle tone and alertness.

Adenosine biosynthesis and metabolism

Adenosine is a purine nucleoside formed from the hydrolysis of ATP, mainly in the glial cells of the brain. ATP is degraded to adenosine by enzymes such as ectonucleotidase and ecto-5'-nucleotidase. Once formed, adenosine is released into the extracellular space, where it exerts its effects through activation of specific receptors, principally A1 and A2A.

A1 receptors are involved in promoting sleep by inhibiting neuronal activity, while A2A receptors promote alertness by facilitating the release of neurotransmitters such as dopamine. Adenosine is then rapidly recycled by specialized transporters or metabolized to inosine by adenosine deaminase.

Therapeutic perspectives

A detailed understanding of orexin and adenosine biosynthesis and metabolism paves the way for the development of novel therapies targeting these systems for the treatment of sleep disorders and other neurological conditions. Pharmacological approaches aimed at modulating orexin or adenosine levels, or influencing their receptors, could potentially restore proper regulation of the sleep-wake cycle in patients with narcolepsy, insomnia or other related diseases.

Section 2.2: Receptors and signalling pathways involved in their action

The biological effects of orexin and adenosine are mediated by the activation of specific receptors and complex intracellular signalling pathways. These receptors, present on the surface of nerve cells and in other tissues, play a crucial role in signal transmission and the modulation of essential physiological processes.

Orexin receptors :

Orexin acts primarily via two types of receptor, Orexin-A and Orexin-B (OX1R and OX2R), which are members of the G protein-coupled receptor (GPCR) family. When orexin binds to these receptors, it triggers an intracellular signaling cascade involving protein kinases and other mediators that directly influence alertness and sleep-wake cycles.

Adenosine receptors:

Adenosine exerts its effects via several receptor subtypes, including A1, A2A, A2B and A3, which are also G protein-coupled receptors. These receptors are widely distributed in the brain and body, and regulate a range of biological processes, including the modulation of alertness and the response to metabolic stress.

Intracellular signaling pathways: Once activated, orexin and adenosine receptors activate various intracellular signaling pathways. These pathways include phosphorylation and dephosphorylation cascades controlled by kinases and phosphatases, as well as post-translational modifications of target proteins. These intracellular signaling mechanisms are crucial for transmitting the effects of neurotransmitters to brain nuclei and other cellular targets, thereby modulating neuronal activity and physiological response.

Therapeutic implications:

Understanding these molecular and cellular mechanisms in detail offers valuable opportunities for developing new targeted therapies. By selectively targeting specific orexin or adenosine receptors or modulating their signaling pathways, it becomes possible to optimize treatments for sleep disorders and other conditions associated with disturbances of these neurotransmitters. This promising approach could pave the way for more effective and better-tolerated therapeutic interventions to improve patients' quality of life.

Section 2.3: Influence of Metals and Minerals on their Functioning

As a bioinorganist, exploring the influence of metals and minerals on orexin and adenosine function can provide unique insights into the underlying biochemical interactions. This perspective may open up new avenues of research to shed light on the still poorly understood mechanisms of sleep disorders.

Influence of Metals and Minerals on their Functioning

As a bioinorganist, exploring the influence of metals and minerals on orexin and adenosine function offers a novel perspective for understanding the biochemical interactions crucial in the regulation of sleep and wakefulness. Metals and minerals can act in several ways to modulate the activity of these neurotransmitters, thereby influencing the physiological processes associated with sleep disorders.

Metallic interactions with Orexine and Adenosine

Metals can bind directly to orexin and adenosine, affecting their three-dimensional structure and functional activity. For example, metals such as zinc, iron and copper are known to bind to specific sites in proteins, modulating their stability and function. This metal interaction can influence the synthesis, release or degradation of orexin and adenosine in the brain, regulating their availability and efficiency in neuronal transmission.

Impact of Minerals on Neuronal Signaling

In addition to metals, certain minerals such as magnesium and calcium play a crucial role in neuronal signalling involved in the regulation of sleep and wakefulness. Magnesium, for example, is essential for the stabilization of cell membranes and the regulation of ion channels, which can influence the response of neurons to neurotransmitters such as orexin and adenosine. Similarly, calcium is involved in the release of pre-synaptic neurotransmitters and the modulation of post-synaptic receptors, thus affecting neuronal signal transmission associated with the sleep-wake cycle.

Potential Applications in Research and Therapeutics

The identification of specific interactions between metals, minerals and neurotransmitters such as orexin and adenosine opens up new avenues for research into sleep disorders. By better understanding these biochemical mechanisms, it may be possible to develop innovative therapies that specifically target the deficiencies or dysfunctions associated with these metal-mineral interactions. Such advances could lead to more effective and better-targeted treatments for patients suffering from narcolepsy, insomnia and other sleep disorders.

Chapter 3: Clinical implications

Section 3.1: Neurological diseases linked to Orexine and Adenosine Dysfunction

Narcolepsy-cataplexy and other sleep disorders are profoundly influenced by orexin and adenosine dysfunction. Understanding these links is crucial to apprehending the clinical variability of symptoms and developing personalized approaches to medical management.

Concrete examples and methodologies used

Narcolepsy-cataplexy: Narcolepsy-cataplexy is characterized by decreased or absent orexin production in the brain, due to autoimmune destruction of orexin-producing neurons. Research methods include brain imaging to study orexin levels and correlations with clinical symptoms in patients. Genetic and immunological studies are also being carried out to identify predisposing factors and mechanisms underlying this dysfunction.

Insomnia: Although less studied than narcolepsy, insomnia can also be influenced by dysfunctions in adenosine regulation. Pharmacological approaches targeting adenosine receptors, as well as studies of adenosine metabolism in the brain, have been used to explore the mechanisms of this condition and develop new therapies.

Section 3.2: Current therapeutic approaches targeting these systems

Current therapeutic approaches to modulate orexin and adenosine activity include the use of norepinephrine reuptake inhibitors and other pharmacological agents. Evaluating their long-term efficacy and safety is essential to optimize symptom management in patients.

Concrete examples and evaluations

Modafinil and other stimulants: Modafinil is a stimulant prescribed to treat excessive daytime sleepiness associated with narcolepsy. It works by increasing the release of neurotransmitters such as dopamine and noradrenaline, thus improving alertness without disturbing nocturnal sleep. Clinical trials have demonstrated its short-term efficacy, but longitudinal studies are needed to assess its long-term impact on patients' cognitive function and well-being.

Agomelatine: This drug acts as a melatonin receptor agonist and serotonin receptor antagonist, influencing sleep-wake cycles. Although initially used to treat depression, its potential to regulate circadian regulation could also benefit patients suffering from sleep disorders associated with abnormalities in adenosine signaling.

By integrating these innovative pharmacological approaches, research aims to improve people's quality of life by optimizing treatments for complex sleep disorders and exploring new therapeutic avenues based on an in-depth understanding of the neurochemical systems involved.

Section 3.3: Potential use in other neurological and psychiatric conditions

In addition to sleep disorders, exploring the effects of orexin and adenosine offers new perspectives for the treatment of complex neurological and psychiatric conditions. This broader perspective opens the way to new therapeutic applications.

Exploring New Therapeutic Applications

Depression and anxiety: Dysfunctions in the regulation of orexin and adenosine have been associated with symptoms of depression and anxiety. Research into modulators of the orexin-adenosine systems could lead to the development of more effective treatments for these conditions, by directly targeting the mechanisms underlying their pathogenesis.

Neurodegenerative diseases: Orexine and Adenosine dysfunction

Recent studies suggest links between orexin and adenosine dysfunction and certain neurodegenerative diseases such as Alzheimer's and Parkinson's. Understanding these links may open up new avenues for the development of neuroprotective therapies aimed at slowing the progression of these conditions by modulating affected neuronal signaling. Understanding these links may open up new avenues for the development of neuroprotective therapies aimed at slowing the progression of these conditions by modulating affected neuronal signaling.

Definition of neurodegenerative diseases

Alzheimer's disease: Alzheimer's disease is a progressive, irreversible neurodegenerative disease of the brain. It is characterized by memory loss, cognitive impairment and behavioral changes. Symptoms generally begin slowly and progressively worsen, affecting the person's ability to carry out daily activities.

Parkinson's disease: Parkinson's disease is a neurodegenerative disorder that mainly affects movement. It is caused by the degeneration of dopamine-producing neurons in a specific region of the brain called the substantia nigra. Symptoms include tremors, muscle stiffness, slowness of movement and balance problems.

Impact of Orexine and Adenosine on these diseases

Orexin (Hypocretin): Orexin is an essential neurotransmitter produced by a small group of neurons in the hypothalamus. It plays a crucial role in regulating sleep, wakefulness and appetite. Research suggests that orexin may also be involved in modulating neuroinflammation and regulating the immune response in the brain. Orexin deficiency may contribute to the pathogenesis of Alzheimer's and Parkinson's disease by affecting these processes.

Adenosine: Adenosine is a neuromodulator that regulates various aspects of brain function, including sleep, wakefulness and stress response. Dysfunctions in the adenosinergic system have been associated with neurodegenerative conditions. For example, studies indicate that adenosine can influence neuroprotection by modulating microglia activity and regulating the inflammatory response, processes crucial in the pathogenesis of Alzheimer's and Parkinson's.

Potential mechanisms and therapeutic prospects

Identifying the links between orexin, adenosine and neurodegenerative diseases opens the way to new therapeutic approaches. For example, strategies aimed at stimulating orexin production or modulating adenosinergic receptors could potentially slow disease progression by attenuating neuroinflammation, improving synaptic function or promoting neuroprotection.

Chapter 4: Technology and advanced research

Section 4.1: Imaging and measurement methods to study Orexine and Adenosine activity

Contemporary brain imaging methods play a crucial role in the exploration of orexin and adenosine, offering direct visualization of their distribution and activity in the human and animal brain. Functional Magnetic Resonance Imaging (fMRI) stands out for its high spatial and temporal resolution, enabling the mapping of brain regions activated during the regulation of sleep and wakefulness. For example, studies have revealed that orexin-rich hypothalamic regions are active during wakefulness and inhibited during sleep, highlighting their crucial role in modulating the state of alertness.

To complement these advances, molecular imaging, such as positron emission tomography (PET), offers the possibility of monitoring the metabolic activity and expression of adenosine receptors in the brain. This approach is of key importance for understanding how adenosine accumulation, in response to fatigue or in anticipation of sleep, alters neuronal signaling and influences the state of alertness. These emerging technologies provide crucial data for identifying neurological disturbances associated with sleep disorders such as narcolepsy-cataplexy and insomnia.

By integrating these advanced imaging methods, research is advancing our understanding of the complex interactions between orexin, adenosine and the regulation of wakefulness and sleep. These approaches open up new prospects for developing innovative diagnostic and therapeutic strategies aimed at improving the quality of life of individuals affected by these neurological disorders.

Section 4.2: Experimental models and pharmacological approaches

Animal models play an essential role in simulating pathological conditions linked to orexin and adenosine dysfunction, as well as in assessing the efficacy of potential pharmacological treatments. For example, orexin knockout mice have revealed the crucial importance of this neurotransmitter in the regulation of sleep and wakefulness. These experimental models can also be used to test pharmacological compounds aimed at selectively modulating adenosine receptors, opening up new prospects for the development of innovative therapies.

Advanced pharmacological approaches include the use of specific noradrenaline reuptake inhibitors and other agents modulating the orexinergic and adenosinergic systems. These studies are of vital importance for understanding the impact of these agents on the regulation of wakefulness, and for identifying potential therapeutic targets in patients with chronic sleep disorders.

By integrating these sophisticated experimental models and innovative pharmacological approaches, research aims to deepen our understanding of the neurochemical mechanisms underlying sleep disorders and open up new avenues towards more effective and better-targeted therapeutic interventions. These efforts promise to significantly improve the clinical management of patients and address unmet needs in the complex field of sleep/wake regulation.

Section 4.3: Recent innovations and future developments

Recent technological advances, such as the use of CRISPR-Cas9 for precise genetic modification, are opening up unprecedented perspectives in the study of the molecular mechanisms of orexin and adenosine. Protein engineering enables the creation of specific receptor variants, facilitating a detailed understanding of their function and paving the way for personalized therapeutic approaches. In parallel, gene therapy explores the possibility of restoring or modulating orexin and adenosine expression in animal models, with potential implications for future clinical applications.

These innovations have the potential to fundamentally transform the treatment of sleep disorders by enabling more precise and targeted intervention on the neurochemical pathways regulating the state of alertness. By harnessing these advanced technologies, researchers aim to develop innovative therapeutic strategies that could not only improve sleep quality, but also effectively treat the complex disorders associated with the regulation of wakefulness and sleep.

This multidisciplinary approach promises to push back the current boundaries of neuroscience research and catalyze the development of new therapies to improve the health and well-being of individuals affected by sleep disorders and other related neurological conditions.

Chapter 5: Perspectives on Vigilance Research

Section 5.1: Complex interactions between adenosine, coenzyme B12 and vigilance

Vigilance, crucial for cognitive and physical functioning, is governed by a complex network of neurotransmitters and biochemical processes. Adenosine, a key neuromodulator involved in the regulation of sleep and wakefulness, plays a central role in the modulation of vigilance. When the brain faces fatigue or prepares for sleep, adenosine accumulation inhibits neuronal activity, signaling biological systems to slow down and rest.

One of the most intriguing implications of adenosine accumulation lies in its impact on the biosynthesis of coenzyme B12, essential for various cellular processes including DNA methylation and nucleic acid synthesis. 5,6-deoxyadenosylcobalamin, the active form of coenzyme B12, is essential for DNA methylation, a process crucial to epigenetic regulation and genomic stability. However, when adenosine accumulates, it can disrupt the availability of substrates required for efficient biosynthesis of this coenzyme. As a result, DNA methylation may be compromised, potentially affecting gene expression and cellular response to environmental and metabolic stresses.

This complex interrelationship between adenosine and coenzyme B12 provides a rich framework for exploring the molecular mechanisms underlying vigilance regulation and genomic integrity. Understanding these interactions not only sheds light on fundamental biological processes, but also opens the door to new therapeutic approaches for treating vigilance disorders and other conditions influenced by these essential biochemical pathways.

Section 5.2: Impact on Cellular Energy and Associated Fatigue

Alongside its influence on coenzyme B12 biosynthesis, adenosine accumulation can also lead to a reduction in the ATP energy available to ribosomes. ATP plays a crucial role in protein translation processes within ribosomes, an essential step in the synthesis of proteins needed to maintain cell structure and function. When ATP becomes limited due to disruption of adenosinergic processes, this can contribute to increased cell fatigue, compromising the ability of cells to maintain optimal functionality and respond effectively to metabolic needs.

This reduction in ATP energy due to adenosine accumulation directly impacts cells' ability to sustain normal metabolic activities and maintain functional integrity. By exploring these mechanisms, it becomes possible to better understand the biological underpinnings of cell fatigue, and consider strategies to mitigate its negative effects on cellular health and general well-being.

Section 5.3: Prospects for research and development of therapies

Understanding the complex interactions between adenosine, coenzyme B12 biosynthesis and cellular energy opens up new avenues for research into alertness and fatigue management. Innovative approaches could aim to selectively modulate adenosinergic pathways to restore normal regulation of sleep and wakefulness, while optimizing the availability of coenzyme B12 and cellular energy. Further studies are needed to elucidate the mechanisms underlying these interactions and to develop therapeutic strategies that could potentially improve alertness and reduce fatigue associated with conditions such as narcolepsy, insomnia and other sleep disorders.

Section 5.4: Advanced technologies and methodological approaches

The rapid development of functional neuroimaging technologies and in vitro and in vivo experimental models represents a significant advance in the study of complex molecular and cellular interactions. These advances offer valuable tools for exploring the effects of adenosine on key biological processes, such as coenzyme B12 biosynthesis and cellular energy dynamics.

Advanced Functional Neuroimaging: Nuclear Magnetic Resonance (NMR) spectroscopy enables detailed visualization of metabolites and biochemical processes in the brain. For example, it can quantify coenzyme B12 levels in different brain regions and study how adenosine alters these levels in response to specific stimuli.

Fluorescence microscopy: This technique enables cellular processes to be visualized and monitored in real time. Using specific fluorescent markers, researchers can observe how adenosine influences energy dynamics at the cellular level, notably by examining changes in amino acid and fatty acid metabolism, essential processes regulated by coenzyme B12.

Genetically modified animal models: Genetically modified animal models, such as knockout mice for adenosine receptors or for key enzymes involved in coenzyme B12 biosynthesis, help to unravel the specific roles of these molecules in the regulation of alertness and fatigue. These models provide valuable insights into the underlying molecular mechanisms, and can serve as platforms for testing new interventions.

Therapeutic Implications and Mechanistic Understanding: These advanced methodological approaches are crucial for obtaining precise and detailed data on adenosine-coenzyme B12 interactions. They facilitate not only the discovery of new targets for vigilance disorders, but also a better understanding of the biological mechanisms underlying complex phenomena such as chronic fatigue and sleep disorders. By integrating these technologies, researchers can develop more targeted and effective strategies, promoting significant advances in biological research and knowledge translation.

This multidimensional, integrative approach provides a robust framework for exploring the nuances of complex biological interactions and **translating this** knowledge into potential applications in neurobiology and basic research.

Conclusion

In summary, orexin and adenosine play crucial but distinct roles in the regulation of sleep and wakefulness, directly influencing neurological health and quality of life. Understanding their complex interplay is essential for developing innovative therapeutic strategies and improving the management of sleep disorders and other neurological conditions. Through a multidisciplinary approach, integrating bioinorganic perspectives with neurology, this book aims to enrich the scientific dialogue on these key neurotransmitters.

This dynamic between activation by orexin and inhibition by adenosine is crucial for maintaining a healthy sleep cycle. Dysfunctions in these systems can lead to disorders such as narcolepsy, where the loss of orexin-producing cells results in excessive daytime sleepiness and episodes of cataplexy.

In terms of research, this understanding paves the way for innovative therapeutic strategies. Current treatments aimed at restoring optimal neurochemical balance use drugs targeting noradrenaline regulation or modulating adenosine receptors. Technological advances in brain imaging, genetic manipulation and pharmacology offer new possibilities for more targeted and personalized interventions.

Taking a multidisciplinary approach, this book merges bioinorganic perspectives with neurology to enrich our understanding of the mechanisms underlying sleep disorders. By integrating insights from metal and mineral biochemistry with neurobiology, it aspires to transform the landscape of scientific research and promote new advances in the management of neurological disease and the promotion of brain health.

In conclusion, this book not only documents the importance of orexin and adenosine in sleep regulation, but also invites reflection on the future of neurological care. Collaboration between different scientific disciplines promises significant advances in the understanding and management of sleep disorders and other complex neurological conditions.

Lexicon :

Orexin (Hypocretin): Neurotransmitter produced in the hypothalamus, involved in regulating sleep, wakefulness and appetite.

Adenosine: Neuromodulator involved in the regulation of sleep, wakefulness and other brain functions. Crucial for the biosynthesis of coenzyme B12, essential for DNA methylation.

Coenzyme B12 (Cobalamin): Essential for DNA and RNA synthesis, as well as amino acid metabolism. Includes 5,6-deoxyadenosylcobalamin, active in DNA methylation.

Narcolepsy: Sleep disorder characterized by excessive sleepiness and sudden sleep episodes.

Cataplexy: Sudden, temporary loss of muscle tone triggered by strong emotions.

Neurotransmitter: Chemical substance enabling the transmission of signals from one neuron to another via synapses.

Neurodegenerative: Refers to conditions where brain neurons undergo progressive degeneration.

Neuroinflammation: Inflammatory response in the brain, involving immune cells such as microglia.

Neuroprotection: Strategies to protect neurons from damage and death.

Pathogenesis: Process by which a disease develops and progresses in the body.

REM sleep: Phase of sleep when dreams occur, associated with intense brain activity.

Polysomnography: A test used to evaluate brain activity and other bodily functions during sleep.

Receptor: Cellular structure that specifically recognizes and binds to substances such as neurotransmitters.

Microglia: A type of immune cell in the brain that plays a key role in the inflammatory response.

Vigilance: A state of watchfulness and ability to stay awake and alert.

ATP (Adenosine triphosphate): Main cellular energy carrier, necessary for many biological processes, including protein translation in ribosomes.

Ribosome: Cellular organelles responsible for protein synthesis from mRNA.

Brain imaging: Techniques for visualizing brain structures and functions, such as functional magnetic resonance imaging (fMRI) and positron emission tomography (PET).

Animal models: Organisms used to simulate pathological conditions in humans in order to study biological mechanisms and evaluate potential treatments.

Knockout mice: genetically modified animal models in which a specific gene is inactivated to study the effects of its absence on the phenotype.

Noradrenaline reuptake inhibitors: Drugs that block the reuptake of noradrenaline in the brain, thereby increasing its synaptic concentration to improve wakefulness and alertness.

CRISPR-Cas9: Genetic modification tool based on a molecular biology technology that enables DNA sequences to be precisely targeted and modified.

Protein engineering: the process of modifying or creating specific proteins by genetic manipulation to study their structure and function.

Gene therapy: Therapeutic approach aimed at introducing functional genes into cells to treat genetic or acquired diseases.

Precise genetic manipulation: using techniques like CRISPR-Cas9 to specifically modify genes with high precision.

Receptor variants: modified forms of cellular receptors designed to study their functional activity in response to different stimuli.

Neurochemical pathways: Neural communication networks that regulate various biological processes, including the regulation of alertness.

References :

Nishino, S., & Mignot, E. (1997). Narcolepsy and cataplexy. In J. M. S. Pearce (Ed.), Neurological Disorders (pp. 1-25). New York: Springer.

Porkka-Heiskanen, T., & Kalinchuk, A. V. (2011). Adenosine, energy metabolism and sleep homeostasis. Sleep Medicine Reviews, 15(2), 123-135. doi:10.1016/j.smrv.2010.06.002

Sakurai, T. (2007). The neural circuit of orexin (hypocretin): Maintaining sleep and wakefulness. Nature Reviews Neuroscience, 8(3), 171-181. doi:10.1038/nrn2092

St Croix, C. M., & Wasserloos, K. J. (2018). Metal-binding properties and structural stability of hypocretin peptides. BioMetals, 31(5), 787-798. doi:10.1007/s10534-018-0146-9

Młyniec, K., Davies, C. L., de Agüero Sánchez, I. G., Pytka, K., Budziszewska, B., Nowak, G., & Time, M. T. (2014). Essential elements in depression and anxiety. Part I. Pharmacological Reports, 66(4), 534-544. doi:10.1016/j.pharep.2014.03.001

McCord, M. C., & Aizenman, E. (2013). Convergent Ca2+ and Zn2+ signaling regulates apoptotic Kv2.1 K+ currents. Proceedings of the National Academy of Sciences of the United States of America, 110(36), 13988-13993. doi:10.1073/pnas.1308204110

Liguori, C., Placidi, F., Albanese, M., Nuccetelli, M., Izzi, F., Marciani, M. G., & Mercuri, N. B. (2012). CSF β-amyloid levels are altered in narcolepsy: A link with orexin peptides? Sleep, 35(4), 537-541. doi:10.5665/sleep.1742

Lazarus, M., & Oishi, Y. (2018). The role of adenosine in Alzheimer's disease. Current Neuropharmacology, 16(7), 1-9. doi:10.2174/1570159X15666171012103514

Jenner, P. (2003). Oxidative stress in Parkinson's disease. Annals of Neurology, 53(Suppl 3), S26-S36. doi:10.1002/ana.10483

Lasek, A. W., Gesch, J., Giorgetti, A., Kharazia, V., & Heberlein, U. (2011). Altered sleep regulation in Drosophila mutants with reduced excitability or hyperexcitability. Journal of Neuroscience, 31(29), 10437-10448. doi:10.1523/JNEUROSCI.0294-11.2011. PMID: 21775597; PMCID: PMC3151916

Chemelli, R. M., Willie, J. T., Sinton, C. M., Elmquist, J. K., Scammell, T., Lee, C., ... Yanagisawa, M. (1999). Narcolepsy in orexin knockout mice: molecular genetics of sleep regulation. Cell, 98(4), 437-451. doi:10.1016/s0092-8674(00)81973-x. PMID: 10481909.

Halassa, M. M., Florian, C., Fellin, T., Munoz, J. R., Lee, S. Y., Abel, T., ... Frank, M. G. (2009). Astrocytic modulation of sleep homeostasis and cognitive consequences of sleep loss. Neuron, 63(6), 865-876. doi:10.1016/j.neuron.2009.08.034. PMID: 19874788; PMCID: PMC2782887.

Borbély, A. A., Tobler, I., & Hanagasioglu, M. (1984). Effect of sleep deprivation on sleep and EEG power spectra in the rat. Behavioural Brain Research, 14(3), 171-182. doi:10.1016/0166-4328(84)90187-9. PMID: 6707028.

I want morebooks!

Buy your books fast and straightforward online - at one of world's fastest growing online book stores! Environmentally sound due to Print-on-Demand technologies.

Buy your books online at
www.morebooks.shop

Kaufen Sie Ihre Bücher schnell und unkompliziert online – auf einer der am schnellsten wachsenden Buchhandelsplattformen weltweit! Dank Print-On-Demand umwelt- und ressourcenschonend produziert.

Bücher schneller online kaufen
www.morebooks.shop

Printed by Books on Demand GmbH, Norderstedt / Germany